I0812676

The INSIDE GUIDE

STEM CAREERS

Jobs in Genetic Engineering

By Beatrice Harris

Cavendish Square

Published in 2024 by Cavendish Square Publishing, LLC
2544 Clinton Street Buffalo, NY 14224

Copyright © 2024 by Cavendish Square Publishing, LLC

First Edition

No part of this publication may be reproduced, stored in a retrieval system, or transmitted in any form or by any means—electronic, mechanical, photocopying, recording, or otherwise—without the prior permission of the copyright owner. Request for permission should be addressed to Permissions, Cavendish Square Publishing, 2544 Clinton Street Buffalo, NY 14224. Tel (877) 980-4450; fax (877) 980-4454.

Website: cavendishsq.com

This publication represents the opinions and views of the author based on their personal experience, knowledge, and research. The information in this book serves as a general guide only. The author and publisher have used their best efforts in preparing this book and disclaim liability rising directly or indirectly from the use and application of this book.

All websites were available and accurate when this book was sent to press.

Library of Congress Cataloging-in-Publication Data

Names: Harris, Beatrice, author.
Title: Jobs in genetic engineering / Beatrice Harris.
Description: New York : Cavendish Square Publishing, [2024] | Series: The inside guide: STEM careers | Includes bibliographical references and index.
Identifiers: LCCN 2023030133 (print) | LCCN 2023030134 (ebook) | ISBN 9781502670342 (library binding) | ISBN 9781502670335 (paperback) | ISBN 9781502670359 (ebook)
Subjects: LCSH: Genetic engineering–Juvenile literature. | Genetic engineering–Vocational guidance–Juvenile literature.
Classification: LCC QH442 .H375 2024 (print) | LCC QH442 (ebook) | DDC 660.6/5–dc23/eng/20230722
LC record available at https://lccn.loc.gov/2023030133
LC ebook record available at https://lccn.loc.gov/2023030134

Editor: Therese Shea
Copyeditor: Shannon Harts
Designer: Deanna Lepovich

The photographs in this book are used by permission and through the courtesy of: Cover, p. 4 Maksim Shmeljov/Shutterstock.com; p. 6 Monkey Business Images/Shutterstock.com; pp. 7 (top), 12 (top) Designua/Shutterstock.com; p. 7 (inset) angellodeco/Shutterstock.com; p. 8 Glynnis Jones/Shutterstock.com; p. 9 Gorodenkoff/Shutterstock.com; p. 10 Microgen/Shutterstock.com; p. 12 (inset) Image Point Fr/Shutterstock.com; p. 14 Daniel Beckemeier/Shutterstock.com; p. 15 (top) BigBearCamera/Shuttersrtock.com; p. 15 (inset left) Jennifer Doudna - 26658740020/Wikimedia Commons; p. 15 (inset right) Emmanuelle Charpentier/Wikimedia Commons; p. 16 PARALAXIS/Shutterstock.com; p. 17 a katz/Shutterstock.com; p. 18 Tomasz Klejdysz/Shutterstock.com; p. 19 Manop Boonpeng/Shutterstock.com; p. 20 TFoxFoto/Shutterstock.com; p. 21 Michael Blann/iStock; p. 22 Dmytro Zinkevych/Shutterstock.com; p. 24 Deemerwha studio/Shutterstock.com; p. 25 (top) Bjoern Wylezich/Shutterstock.com; p. 25 (inset) Vadim Sadovski/Shutterstock.com; p. 26 Caulobacter crescentus/Wikimedia Commons; p. 27 Ground Picture/Shutterstock.com; p. 29 (left) Stock-Asso/Shutterstock.com; p. 29 (right) arindambanerjee/Shutterstock.com.

Some of the images in this book illustrate individuals who are models. The depictions do not imply actual situations or events.

CPSIA compliance information: Batch #CWCSQ24: For further information contact Cavendish Square Publishing LLC at 1-877-980-4450.

Printed in the United States of America

Find us on

CONTENTS

Genetic engineers are scientists who use technology to change the genes of a living thing.

A DIFFERENT KIND OF ENGINEERING

Chapter One

Engineers are a major part of constructing our world. They use math and science to build and improve objects and processes. One kind of engineering—genetic engineering—has to do with a different kind of building material than most engineers use: genes. Scientists who work in genetic engineering alter genetic material in plants, animals, and people for a variety of reasons, such as to treat people's diseases. This is an exciting and growing career path.

First, What Are Genes?

"Genetic" means having to do with genes. A gene is a basic unit of DNA, which stands for deoxyribonucleic acid. DNA carries the information about how an organism looks and how its different parts work. Some call DNA the instructions for how a body works. Each gene, then, would be a single piece of instruction. DNA can be found in almost every cell in an organism.

Fast Fact

Some viruses contain DNA too.

A gene is called a unit of heredity. Heredity is the passing on of traits, or qualities, from parents to their young. Traits include things like hair and eye color in people. Diseases and other conditions can

Fast Fact

There are about 20,000 to 25,000 genes in the human body. More than 99 percent of all genes are exactly the same in all people.

Height is an example of a polygenic trait. You likely won't inherit the exact height of your mother or father. More than one gene in your DNA is at work.

be passed on as well. How do genes do this? They instruct the body to make proteins that carry out specific functions. Some traits are linked to more than one gene. These are called polygenic traits.

Genes are made up of chemicals called nucleotides. A nucleotide has three molecules: a sugar, a **phosphate** group, and one of four nitrogen-containing bases (adenine, guanine, cytosine, or thymine). The nucleotides form two chains that look like a ladder that spirals around. This is the shape of DNA, which is called a double helix.

The DNA in one cell stretched out would be about 6.6 feet (2 meters). However, it's packed tightly into a cell's nucleus in

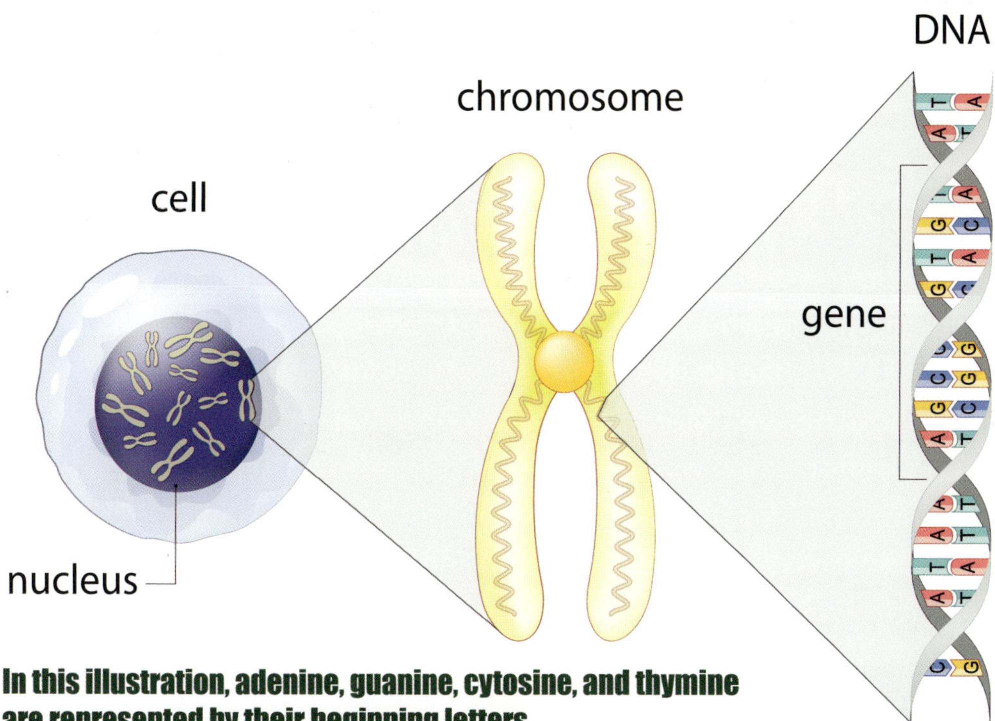

In this illustration, adenine, guanine, cytosine, and thymine are represented by their beginning letters.

Fast Fact

All the DNA in a person's body uncoiled and placed end to end would be 67 billion miles (108 billion kilometers) long!

structures called chromosomes. A cell usually has 23 pairs of chromosomes, or 46 total. (People receive a chromosome from each parent to make a pair.)

An entire set of DNA is called a genome. The mapping of the human genome was completed in 2022. Scientists used a number of people's DNA to map the genes.

THE ETHICS OF GENETIC ENGINEERING

Some people wonder if genetic engineering is ethical. Ethics have to do with morals, or whether something is correct or good. Some people wonder if DNA *should* be changed and worry how such technology can change nature. Genetic engineering has helped some plants become better crops by making them larger or disease resistant, for example. It has also given people hope of eradicating, or ending, certain diseases. However, some wonder if genetic engineering might go too far. For example, should people be able to choose traits for their children? What about adding human DNA to a mouse? (This was an actual failed experiment, which caused mice to develop cancer.)

Some people are against genetically modified organisms (GMOs), especially crops. They worry GMOs used as foods are bad for people's health as well as for the **environment**, though most research suggests these crops are safe to eat.

Cutting-Edge Careers

Clearly, people with careers in genetic engineering know a lot about genes, DNA, and genetics. They may work in fields relating to health, plants, animals, or other organisms. Many aren't called genetic engineers, but have different job titles, as you'll read.

To prepare for the work described in this book, people usually study biology, biochemistry, molecular biology, or similar courses in college. Others may study engineering too. Many workplaces require a master's degree for some lab positions. However, those hoping to go into research or to become leaders in their field get a doctorate, or PhD.

People hoping to pursue a career in a genetic engineering field need to learn exactly where genes are located in an organism's DNA and what they do.

Lab technicians are a critical part of all kinds of genetic engineering. They remove DNA from cells, run experiments, and help in research. Many work in the medical field.

Chapter Two

GENOMIC MEDICINE

Some genetic engineers focus on human DNA, specifically to improve people's health. This is a field called genomic medicine. Doctors, nurses, lab technicians, researchers, and others may practice genomic medicine. Many get a degree in their medical field first and then get additional education and experience in genetics.

Gene Therapy

Clinical geneticists are doctors who specialize in genes and heredity and treat certain genetic conditions. They're involved in both caring for patients and genetic research to help them. Some genetic research has resulted in effective medical treatments called gene therapy.

Gene therapy may mean replacing the gene that causes a disease with a healthy gene or adding a missing gene to a cell. A gene can also be "silenced." This means temporarily blocking a certain gene's message that can have a harmful effect on the body.

As of June 2023, the U.S. Food and Drug Administration (FDA) has approved gene therapies for just a few conditions, including Leber congenital amaurosis, which leads to blindness; a skin cancer called melanoma; a muscle condition called spinal muscular atrophy; and blood cancers including a kind of leukemia and lymphoma.

INSULIN: A SUCCESS STORY

Insulin is a **hormone** that controls the level of glucose (sugar) in the blood. It makes sure some glucose enters cells for energy and some is stored in the liver for later use. People with type 1 diabetes have bodies that can't make insulin or make a limited amount. Genetically engineered insulin using a technology called recombinant DNA has proven to be helpful. A piece of DNA called a plasmid is removed from a bacteria or yeast cell. The gene for human insulin is inserted into the plasmid. The plasmid is then added to a new bacterium or yeast, which makes insulin that can be used for medicine.

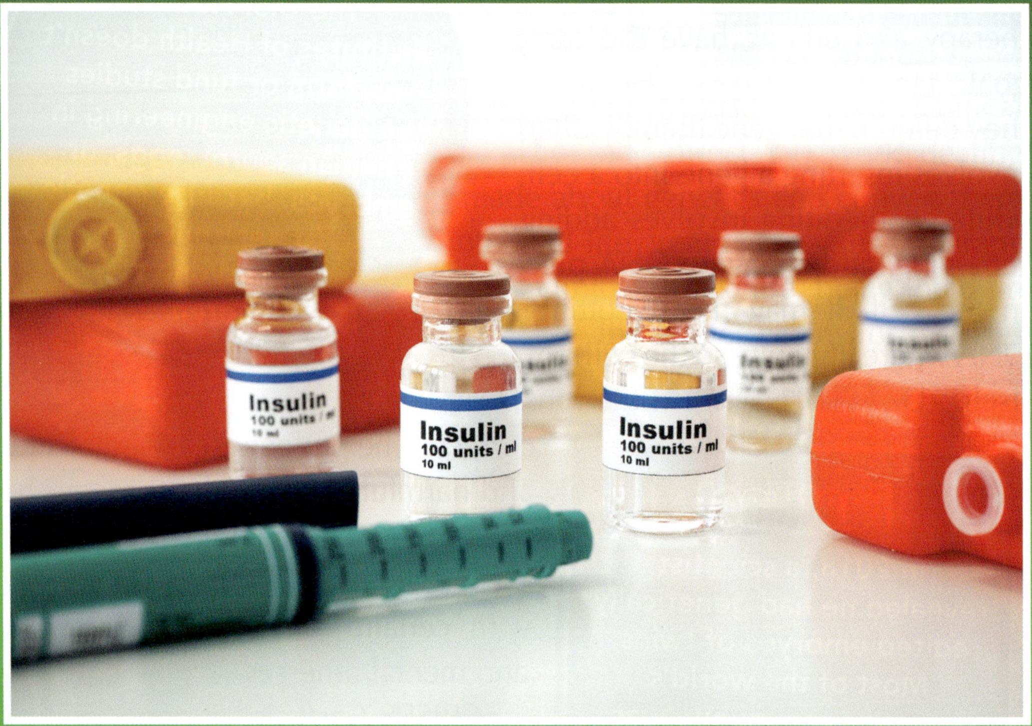

Before man-made insulin, people used cow and pig insulin. In 1978, scientists discovered how to make human-made insulin using bacteria.

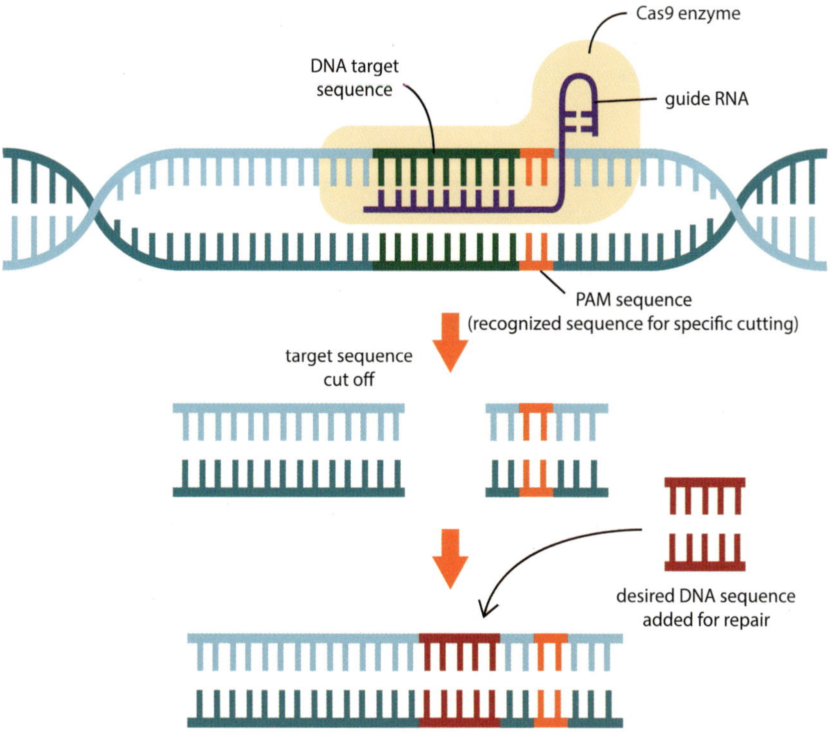

CRISPR-Cas9 technology depends on an enzyme called Cas9 and guide molecules of RNA, which is a natural chemical compound that carries information. Guide RNA leads the enzyme to the right place in the DNA so Cas9 can cut it, preparing it for genetic changes.

medical world. It was inspired by the **immune** response of bacteria. When infected with a virus, bacteria can cut a section of a virus's DNA and put it in their own DNA. CRISPR-Cas9 treatments for people are still in the research stage as of summer 2023.

Jennifer Doudna (*left*), a biochemist, and Emmanuelle Charpentier (*right*), a microbiologist, geneticist, and biochemist, received the 2020 Nobel Prize in Chemistry for the discovery of CRISPR-Cas9 technology.

GMOs, such as corn, can resist pests.

ENGINEERING PLANTS AND ANIMALS

Chapter Three

Genetically modified (GM) food is a big business. You've likely already eaten some, though much is grown as animal feed. A lot of sugar, soybean, and corn products are made with genetically modified, or genetically engineered, crops. The scientists who make these foods possible are sometimes called biotechnologists.

Biotechnologists who are focused on genetics often have another focus, such as certain plants or animals. They use many special lab tools to isolate, or separate, DNA and perform experiments using advanced computer software.

Fast Fact

As of January 2022, GMOs in the United States must have a label telling people they're made or grown with GMO technology.

Biotechnology in Food

Some scientists use their knowledge of biology and genetic engineering technology to solve problems

GM foods have been around since the 1990s. People fought to have them labeled so they could make decisions about whether to eat them.

This is an ear of corn eaten by a European corn borer. These pests can cause great damage to crops and can be hard to detect.

that arise for farmers. For example, two problems farmers face are pests eating and destroying crops and weeds crowding out crops and taking **nutrients** from soil.

Biotechnologists discovered a gene that makes a protein that kills the larvae of the corn borer insect. This gene was in a soil-dwelling bacterium. They removed the gene from the bacterium's genome and added it to the corn's genome at an early stage of the plant's development. The mature corn plant kills some insects when the insects eat it.

Biotechnologists also found a way to make certain crops resistant to herbicides, or weed-killing chemicals, that many farmers spray on their fields. In the case of the commonly-used herbicide glyphosate, which is sold as Roundup, they found a gene that gave the bacteria called *Agrobacteria* resistance to glyphosate. Adding this gene to crops such as soybeans, cotton, and corn meant that fields of these crops could be sprayed with herbicides, killing the weeds but not killing the crops.

Fast Fact

Since the creation of herbicide-resistant crops, "superweeds" have developed that can resist certain herbicides.

Biotechnologists not only alter the DNA of some plants, but also make sure they grow successfully.

Animal Biotechnology

Biotechnology is also being used to alter the genome of animals. Scientists are looking for ways to make animals resistant to disease and illness, to provide meat and other animal products that are healthier for people to consume, and to lessen some animals' negative impact on the environment.

In 2002, scientists in Japan selected a gene called FAD-2 from spinach and introduced it into **fertilized** pig eggs. The piglets that grew from these eggs reportedly had meat with less unhealthy fats than others raised for pork.

In 2022, the FDA permitted that a kind of GM cattle could be raised for food, a first in the United States. Scientists working for a

GMOs AND THE ENVIRONMENT

Agriculture is a contributor to greenhouse gas emissions. Greenhouse gases, such as methane and carbon dioxide, trap heat in Earth's atmosphere and cause climate change. Agriculture is also a cause of pollution, with chemical fertilizers and other pollutants ending up in rivers and other bodies of water. Biotechnologists search for ways for genetics to help. Some are looking into methods to reduce the amount of methane that cattle release into the atmosphere. Ideas include genetically engineering the grass the cattle eat and altering the genetics of the microorganisms that live within the cattle, which produce the methane.

A Canadian biologist genetically altered a pig to produce less of the chemical phosphorus in its waste. Phosphorus can be a pollutant.

The term "transgenic" is used to describe an animal that has genetic material from a different species.

Fast Fact

Biotechnologists often do their work in labs and offices. They may be associated with colleges and universities, research centers, or food processing factories.

breeding company called Acceligen genetically modified the cattle to grow shorter, slicker hair using CRISPR-Cas9 technology. They expect these cattle to have fewer heat-related problems in warmer climates. (Cattle with heat stress may grow more slowly, make less milk, and have a harder time producing young.)

Scientists examining bacteria may need powerful microscopes. These tools can magnify objects up to 1,000 times their actual size.

BENEFITING FROM BACTERIA

Chapter Four

Some genetic engineers use their knowledge of life science and mechanical, chemical, or electrical engineering to find solutions to real-world engineering problems. They're called biological engineers or bioengineers. They may find work in many fields.

Bioengineers often work with people in other STEM (science, technology, engineering, and math) fields in research and development. In recent years, breakthroughs into the genetics of bacteria have had exciting results for environmental cleanup efforts and many industries.

Beneficial Bacteria

Bioengineers and other scientists have created genetically modified bacteria that clean polluted water at wastewater treatment plants. These bacteria can remove heavy metals, toxins, and other pollutants from water.

> **Fast Fact**
>
> U.S. Department of Energy scientists are researching how to use tiny organisms to clean up **nuclear radiation.**

Bioengineers are also researching GM bacteria that can eat plastic. Plastic is a huge problem. Most of it isn't recycled. It's thrown away, but doesn't break down like organic waste. A lot of plastic waste ends up in places where it harms plants and animals.

A SYNTHETIC BIOLOGIST'S BREAKTHROUGH

Scientists called synthetic biologists stitch together lengths of DNA and then add them into an organism's genome. Synthetic biologist Sara Molinari and a team of researchers genetically modified bacteria called *Caulobacter crescentus* so that they could combine to form "living materials." The bacteria could be made to produce hard materials, which could be used for construction, or soft materials, which could be a filling material. The living material can also repair itself if damaged. The research team was able to grow these materials in about 24 hours using an **incubator**, proteins, and sugars.

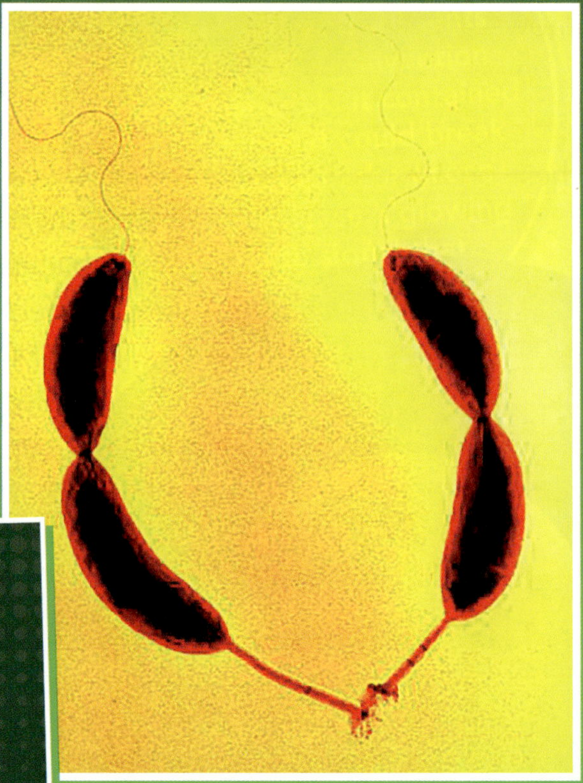

This is a close-up look at *Caulobacter crescentus* bacteria.

Fast Fact

Unlike other kinds of construction materials, the genetically modified *Caulobacter crescentus* can break down easily in nature.

THINK ABOUT IT!

1. Are you fearful of GMOs? Why or why not?
2. Where else could genetic engineering technologies be useful?
3. Do you think genetic engineering should be used to select traits for children in the future? Why or why not?
4. Scientists often test genetic engineering technologies on animals such as mice. Do you think this is ethical? Why or why not?

INDEX